かつじんけん

活人剣

a Sword Arts Journal

Premiere Edition

November 2011

Reprint August 2014

Katsujinken

Meishiha Mugai Ryu

2010 Gasshuku, Jason Hatcher

Bob Elder

A Toyama Ryu Sensei Interview

WMAW

Deed of Arms, Greg Mele & Terry Pfister

Swords & Swordsmen

Mike Loades, Book Re

KATSUJINKEN MAGAZINE

Premiere Edition, Volume 1 * November 2011

Reprint by popular demand August 2014
ww.donnaink.org

FOUNDER | EDITOR-IN-CHIEF
Jason Lee A. Hatcher

ADVERTISING TEAM
Jason Lee A. Hatcher
Donna L. Quesinberry

CREATIVE ART DIRECTOR
Kent Jensen

EDITORIAL TEAM
Jason Lee A. Hatcher
Kent Jensen
Donna L. Quesinberry

PUBLISHER
DonnaInk Publications, L.L.C.

SUBSCRIPTIONS | PRO-SHOP
Michelle R. Hatcher

ADVERTISERS

Akido in Fredericksburg, Aviv Dojo, Capital Area Budokai, Chicago Swordplay Guild, dpInk: DonnaInk Publications, East Coast Martial Arts Supplies, Katsujinken Magazine, Mugendo Budogu, Nihonzashi, Purple Heart Armory, Revival Clothing, Senpokan, Becky Sheetz-Runkle, Shining Moon, Swords and Swordsmen, Sword Forum International, Tameshigiri Mats, The Japanese Sword, United States Batto Do Federation, US Hombu Dojo, Virginia Kenshinkai.

CONTRIBUTORS

Bob Elder, David Drawdy, Jason Hatcher, Carl Jenkins, Kent Jensen, Sang Kim, Mike Loades, Lorenzo Macuja, Paul Martin, Greg Mele, Richard McCluer, Dave Neeley, Victor Herrero Perez, Terry Pfister, Becky Sheetz-Runkle, Christopher Treichel, Doug Waller, Robert Zingg.

PHOTOGRAPHY AND IMAGES

David Zlotin of DEOPages, Brenda and Tom Turk of Piratical Photography, Jim Blocker of Middle Blocker Photography—see the Swordfest Pictorial for more information.

ADVERTISE WITH US: contact Jason Lee A. Hatcher at (301) 356-6552 or email samuraihitachi@yahoo.com.

SUBSCRIBE: contact Michelle R. Hatcher, her email is michelle@katsujinkenmagazine.com. Limited Edition print journal, published four times a year. Subscriptions available at website or mailing address as listed. Individual Limited Edition copy to 1,000 sold in US dollars plus appropriate postage for destination. Order information

WRITE FOR US: interested in being an author with Katsujinken? Please contact Katsujinken Magazine and email submit@donnaink.org.

KATSUJINKEN MAGAZINE, A SWORD ARTS JOURNAL
4405 S. Kirkman Rd., Suite #B208, Orlando, FL 32811
Phone: (888) 564-7741
Facsimile: (703) 373-9552
Visit our website at: www.donnaink.org
ISSN: 2164-2672 (Online) | 2164-2664 (Print)

In This Issue

活人剣

"Perceive that which cannot be seen with the eye."
~ Miyamoto Musashi

www.katsujinkenmagazine.com

SENPOKAN
"less chat, more mat"
www.senpokan.com

EDITOR'S MESSAGE

Welcome to the long awaited launch of *Katsujinken Magazine*, your premiere Sword Arts Journal.

I'd like to start by offering my tremendous appreciation to everyone involved in this publication and its growth. I can't begin to express how very fortunate we have been with the influences of so many individuals. Everyone who has helped out has provided phenomenal support to aid this publication. So thank you "all" for all the hard work and consideration.

From the moment I conceived of the idea of building a Sword Arts magazine, I immediately knew the Japanese word Katsujinken, meaning, *The Life Giving Sword or Life Sustaining Sword,* was the most appropriate title for this publication's journey. As you'll read through Kent Jensen's—Samurai Speak (Katsujinken's Art Director and a Japanese linguist) lesson in this issue—you will see very clearly how the concept of Katsujinken applies to the entire Sword Community.

To me, and the Katsujinken Family—including all Swordsmen and Swordswomen the sword truly is "life giving."

Our practices with Sword Arts provide us with outlet to live in a medium of personal expression and growth. What we learn, practice and teach is what makes us who we are in our day-to-day life. Moreover, we find our niche' we align with and call it our own. It seems to each of us, the Sword Art we follow, becomes a driving force within our lives—becoming inseparable on many levels.

Our Sword Arts also result in friendships and bonds, which we would never realize without the conquest of our Art Forms and *The Life Giving Swords* or *Life Sustaining Swords* we encounter on our personal journeys of sword discovery. Out of all of this, the most important thing of it all, is the reality we build community with one another—and we grow as a Family of Practitioners—whether East or West Philosophies.

In my humble opinion, this ideal is not only of Japanese origin, but for all of Sword Artisans. For this reason, Katsujinken, a Sword Arts Journal will involve all communities of edged weaponry. Though our familiarity is Japanese Arts we intend to spread knowledge pertaining to all arts and anticipate Katsujinken will be worthy of all Sword Artisans. It is my sincere hope and faith, we will do justice to those who supply content and contributors will grow as we mature as a publication. Please enjoy our Premiere Edition of Katsujinken—thank you for supporting this new path of discovery.

Domo Arigato Gozaimasu
Jason Lee A. Hatcher

かつじんけん
Katsujinken
活人劍
a Sword Arts Journal

Samurai Speak

Understanding Japanese language and culture of the modern swordsmen

By Kent Jensen, Creative Director Katsujinken Magazine

The "hanko" or seal used in the cover art for this magazine takes advantage of the numerous ways to write words in Japanese; Kanji, Romaji, Hiragana and Seal Script. This article examines the word Katsu-Jin-Ken—活人剣, its meaning, origin and the evolution of the characters that make up the word.

In Japanese, there are two expressions: Satsujinken—殺人剣—the sword which kills or takes life and Katsujinken—活人剣—the sword which gives life. Both expressions contain the same character for sword and person, the object towards which the action of the sword is directed. The only difference is in how the sword is used; in one case taking life and in the other giving or saving the life of a person.

The character Katsu—活—is made up of two parts; the left side meaning water and the right side, taken for its phonetic sound, which was originally thought to be the sound of water gushing indicating "lively activity" or "life".

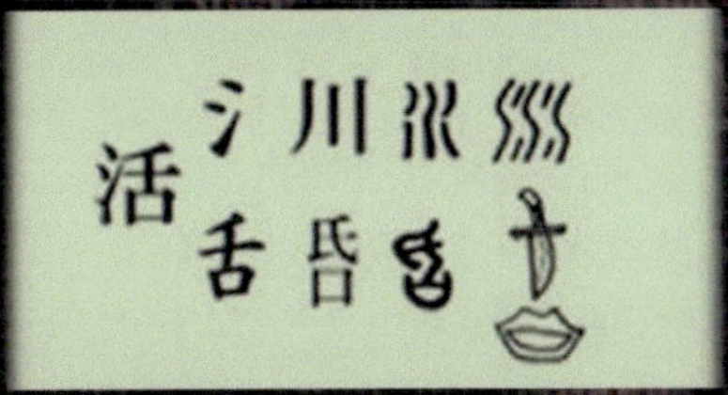

The water element was originally drawn depicting a flowing river which was stylized over time to become 川. When water is used as part of another character the three vertical lines are rotated on their side become two dots flowing diagonally downward and one diagonally upward.

The element on the right side 舌, was originally composed of the pictograph for knife and mouth (the symbol for tongue). Hense, water and tongue form the character for Katsu meaning lively, life or activity.

The character Jin 人, which means "person", is derived from an ancient pictograph depicting a person standing slightly bent over and viewed from the side.

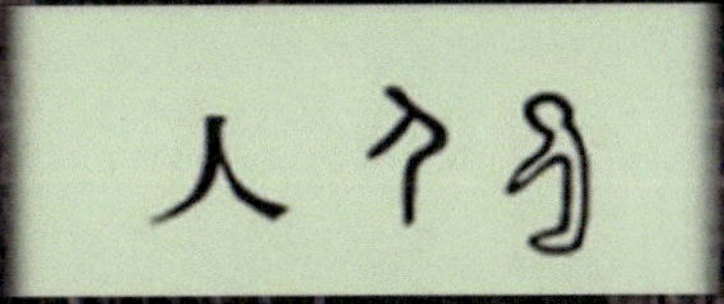

The character Ken— 剣—is composed of an old form element 僉 depicting two boxes or bundles of supplies or goods 口 and two people 人 under a roof, signifying that many things and people were gathered in one place. The right side is a variant of 刀 or sword. The significance of the old form element on the left is unclear. It may have carried the sound in the original Chinese or could have symbolized the clan and its wealth which would be defended by the sword. Regardless of reason it was combined with the variant for sword 刀, which carries the meaning for the character.

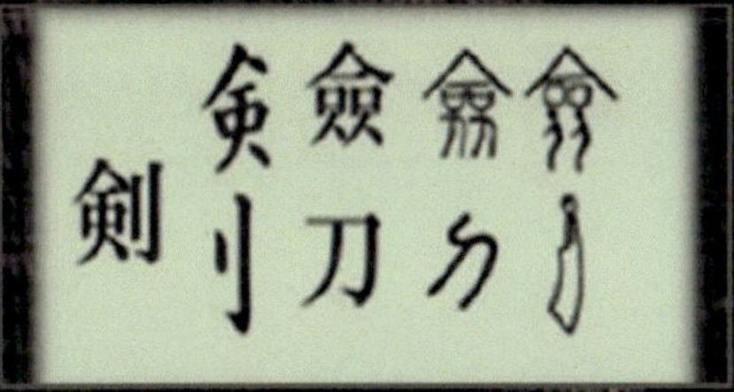

When read together, these three characters take on the meaning of "the sword that gives life to a person"; Katsu—活—life, Jin—人—person, and Ken—剣—sword.

The sword itself represents the warriors' power and is tied to the concepts of obligation and loyalty to superiors. The sword, to the samurai, was an instrument of service and the concept of Katsujinken or giving life by the sword, is intertwined with that philosophy.

活人剣

WHY COMMUNITY IS IMPORTANT

by Lorenzo Macuja

Member of the Byakkokan Dojo New York City, New York

This past weekend, I had the pleasure of attending the 4th annual "Batto Camp" in the Poconos. It was a great weekend, with great training opportunities, and a more relaxed judgment-free environment. Not to mention great food. What struck me most though was the community | family aspect of it. Out of everything I got out of the two days of camp, this was I think the most valuable.

Martial arts in general are a difficult journey, and having a great community atmosphere is, in my mind, essential in maximizing everyone's growth and development. There is the support aspect, where even on your worst days; one can be inspired by your fellow students | sensei. Being encouraged to do your best can help overcome the sometimes self-discouraging feelings one gets from a bad day (or in my case, after recently recovering from an injury).

There is the sharing of information aspect of community. Different viewpoints are always welcome when learning anything. Sometimes a fresh viewpoint is all that's needed to help invigorate your training, or expand on your concepts. Perhaps it just validates what you

Participants from left to right and top to bottom: Ivan Slavov, Ryan Estacio, Lorenzo Macuja, Matt ODell, Robert Crook,, Douglas Firestone, Robert Groce, Neil Earhart, Marky Kurihara, Mayumi Kurihara, Andy Roman, Jason Lee A. Hatcher, Alex O'Dell, Ron Duponte, Sang Kim, Kyle Kim.

Magazine cover inspired by this photo taken at Batto Camp 2010.

know. Especially for beginners, your sensei will be your number one source of information, but feel free to learn from your seniors and other Sensei . . . and discuss what you learn. Maybe you'll get to your own truth...everyone is different, after all.

Another very important thing about having a strong community is that it re-emphasizes the feeling of a greater whole that one should have when doing a martial art. It's not just you, your dojo, and your fellow students. There's a whole lot of other people practicing this art and trying to achieve the next level. Consider that next time you decide to have lazy Kata, or you let your fundamentals slide. People are working hard to preserve a martial art passed on through the generations...why aren't you? Have respect for your community by doing your best always, and keeping standards high. You represent not just your dojo, but the whole community, so always show your best.

Batto camp was a great experience because I finally started to feel like a real part of the community, despite having only been training for a handful of months. It inspired me to try even harder to be a better swordsman, be a better martial artist, and also introduced me to some really great people. This is another set of family members who I know I can count on, and who will also count on me.

We're all working towards the same goals, so why not work on them together?

Sun Tzu for Women

Becky Sheetz-Runkle

The Art of War for Winning in Business is the businesswoman's guide to success, based on the masterpiece of military strategy, The Art of War. Sun Tzu's classic has shaped Eastern military and business thinking. In the West, the book's popularity continues to grow as managers and leaders apply its principles to their business challenges.

Sun Tzu for Women helps women understand, and cultivate their competitive advantages. Many men know this classic well and are using it to be more effective leaders and implementers. But for women, that hasn't been the story. Women have a great deal to learn from Sun Tzu because, largely, women haven't read Sun Tzu. And that's too bad. Because there's awesome potential for today's current and future businesswomen to bring Sun Tzu's lessons to life in practical and actionable ways—without sacrificing the uniquely feminine aspects of their leadership styles.

Sun Tzu for Women shows how to first understand, and then effectively internalize his strategies to advance your career and improve your business. The pace of our lives, our worlds, and our workdays is often so frenetic that true understanding is elusive. The book fosters understanding of Sun Tzu. But more than, it fosters deep understanding of yourself, your conditions and those around you. (And that's really what Sun Tzu had in mind all along.) Victory begins well before the battle is fought. Sun Tzu for Women is for you if you're willing to cultivate within yourself the spirit of self belief required to carry out the tactics of The Art of War.

How Does Any of This Apply to Today? Timeless truths, both simple and complex, never go out of style or function. Sun Tzu for Women uses accomplished business leaders like Carly Fiorina, Meg Whitman, Indra Nooyi, Marsha Serlin and others to illustrate Sun Tzu in practice, Today.

The biggest mistake anyone can make with The Art of War is to deem it too far removed from their challenges to be applicable. Some may believe The Art of War will lead us to seek out combat as our first and best option, that it will train us to follow a "male paradigm" of aggressive leadership. But this is very far from the real Sun Tzu. A concise summary of The Art of War is: He who knows when to fight and

> Timeless truths, both simple and complex, never go out of style or function

when not to fight will win. So it's silly to conclude that we should always be looking for a way to overtake adversaries and colleagues with hostility and treachery. Sun Tzu only advises waging war when there is something to be gained. For most of us, waging war doesn't mean mobilizing men to take a city. It means mobilizing ourselves or our teams to win a big contract, seize a market opportunity, or reposition a company.

Sun Tzu for Women will help you gain a decisive business advantage, in ways least expected: "Take advantage of the enemy's unpreparedness, make your way by unexpected routes."

Becky Sheetz-Runkle

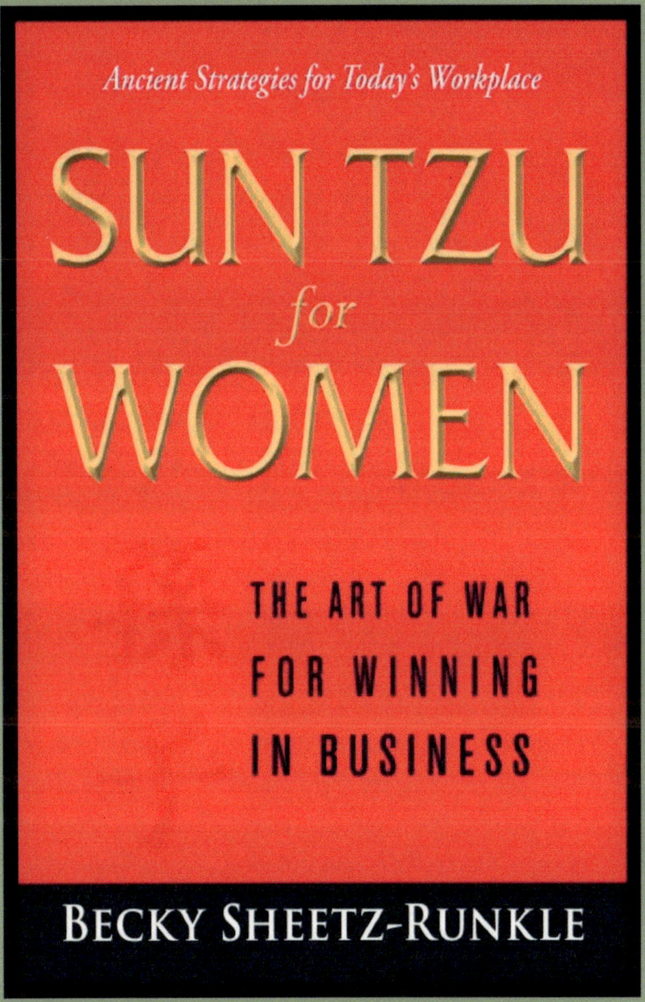

Learn more about Becky at bsheetz@beckysheetz.com.

Follow her on FaceBook (http://www.facebook.com/becky.sheetz) and Twitter (@BeckySheetzSun).

Visit her website today: : www.suntzuforwomen.com

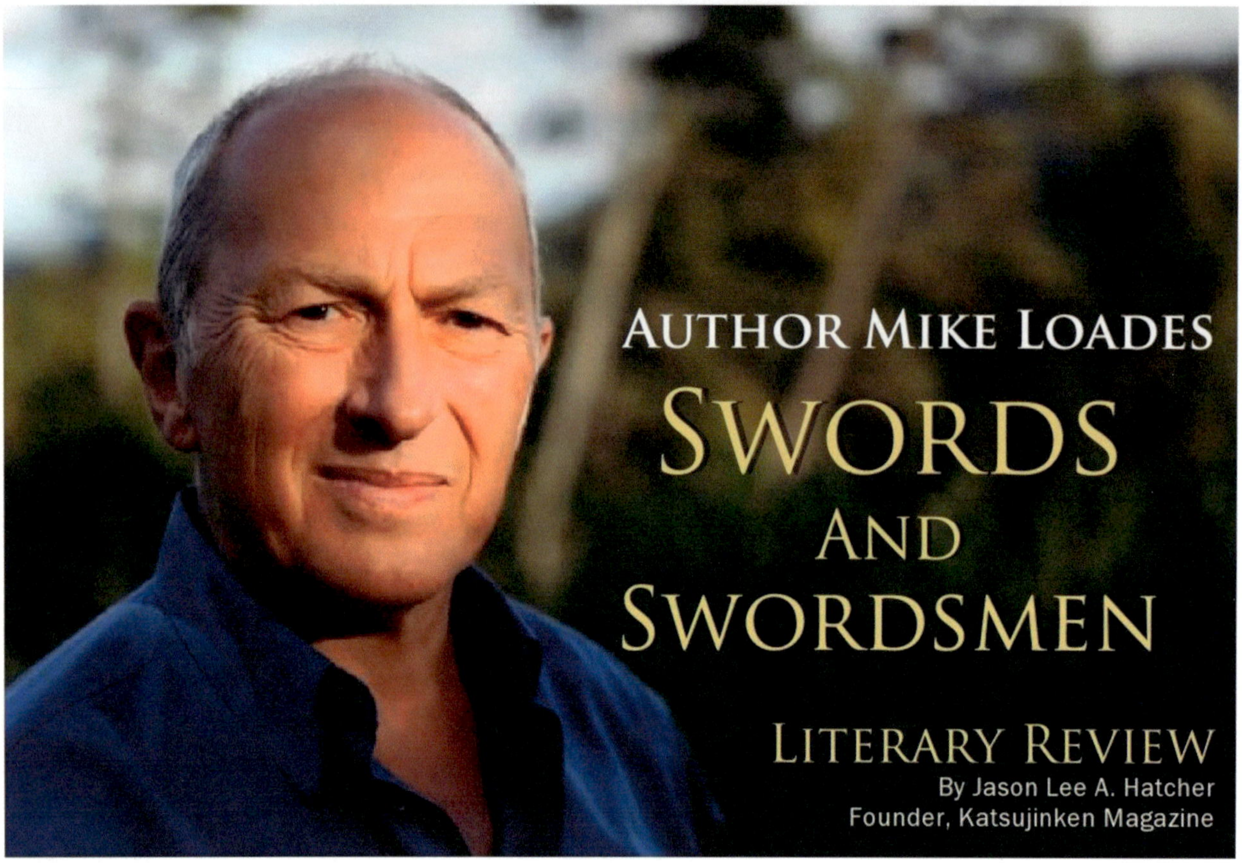

Author Mike Loades
Swords and Swordsmen
Literary Review

By Jason Lee A. Hatcher
Founder, Katsujinken Magazine

From the moment you set your eyes on the cover of Swords and Swordsmen, you get the sneaking suspicion something special is contained within this book's pages. Written by Author, Director, Fight Choreographer, Historical Weapons Expert, and Presenter Mike Loades,' *Swords and Swordsmen* quickly proves to be a satisfying read. Elementally unparalleled, this book is not just for the avid swordsman and swordswoman. Swords and Swordsmen unearths what is inarguably the single most iconic weapon of civilization.

This title supplies every reader with a deeper respect, understanding and conviction regarding 'all things sword.' *Swords and Swordsmen* smoothly progresses from chapter to chapter. There is no feeling of pause a reference text often delivers; no dictionaryesque ambivalence develops. Rather, through his accomplished pen, Mr. Loades commands readers and transports them through an intricately woven fabric of far-reaching sensibilities the sword and its life has provided to humanity throughout recorded history.

Each chapter close features an unfolding in introduction as this captivating memoir of the sword unfurls. Readers experience these artifacts—their weightiness to mankind—while sensing the saged wisdom of the ages. *Swords and Swordsmen* is magical. This is an 'all things sword' ride; readers climb aboard an H. G. Wells Time Machine classic, while Mike charts the course. He delivers his fans to times where swords built nations. He sits us center stage to view the swords' impact on the rise and fall of Great Empires. He sagaciously forges our consciousness in tandem with the swords.'

In his chapter titled, Uesugi Kenshin Samurai Warlord, the Japanese blade is thoroughly examined from creation, to its role in history, to how it relates to Buddhism, to the Bushido Code and how it was employed as well as the blades' use during times of war. This is an informative, succinct, and exceptional book delivering well-paced momentum from start to finish.

Through his accounts of the creation of tamahagane, the precious steel used to forge true Japanese katana, readers feel as if they've been transported to the tatara house itself. The alter-like clay hearth blazes with fiery life combining all atmospheric elements producing new steel. . . Tamahagane. Next, through skillful forging of Swordsmith Ono Yoshimitsu, the birth of a new blade unfolds and the mystery of the swords' creation is revealed.

This work also presents the story of the Samurai sword—delivered with multi-faceted regale. It is in this fashion; Mike Loades presents not only the Japanese blade, but also blades of Celtics, the Great Kings, Greeks, Pharaoh, and Romans as well as all the notable personalities who engaged the sword in their era during their walk in this life.

If you are a sword practitioner—do your library justice—ensure *Swords and Swordsmen* has its place on your book shelf.

This book delivers. This title provides a wealth of information for sword artisans and remains a valuable practice text of today. With rapid expanse of sword art recreation, *Swords and Swordsman* has something for everyone from cover to cover.

It has been my honor to read, review and comment on this work.

Jason Lee A. Hatcher
Founder | President
Katsujinken Magazine

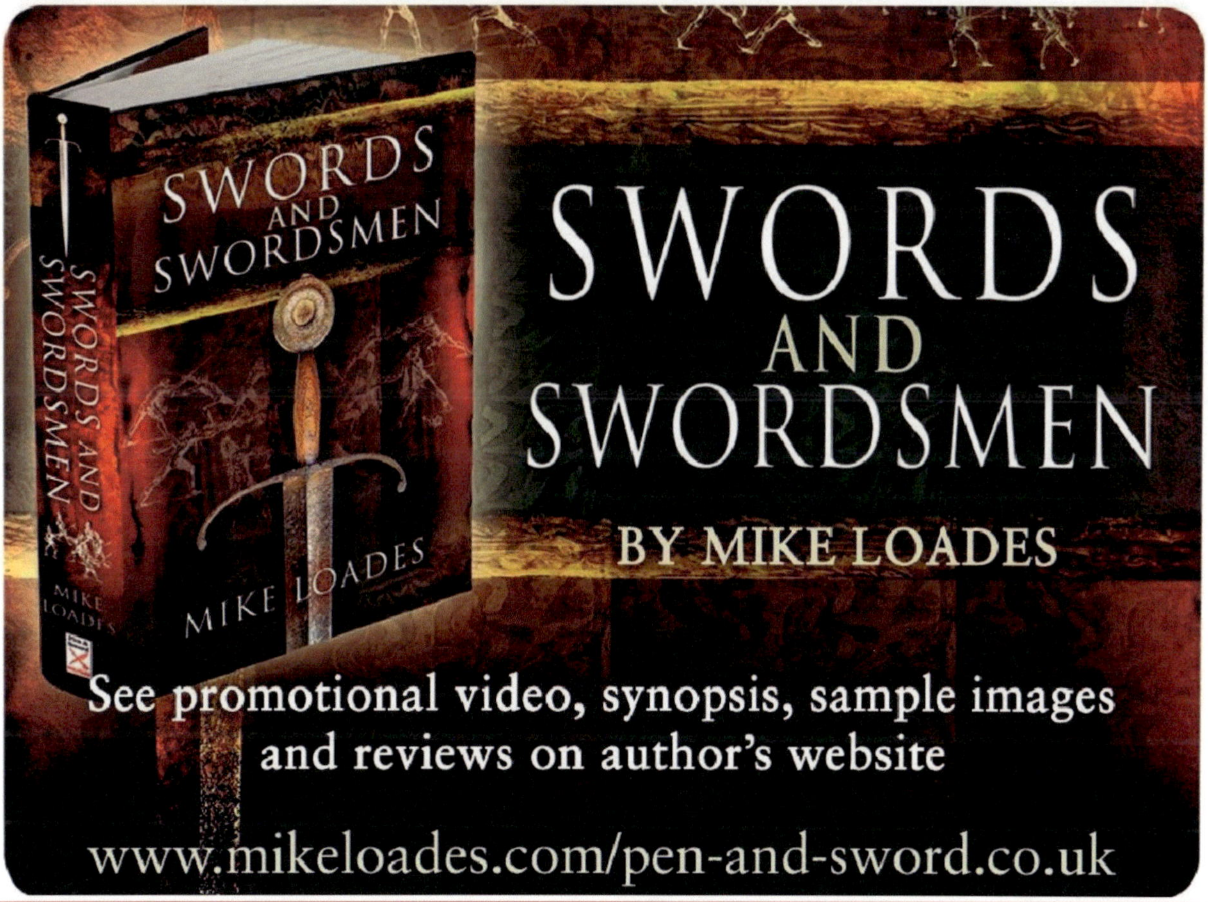

Deed of Arms

WMAW, Western Martial Arts Workshop

By Greg Mele & Terry Pfister
Chicago Swordplay Guild

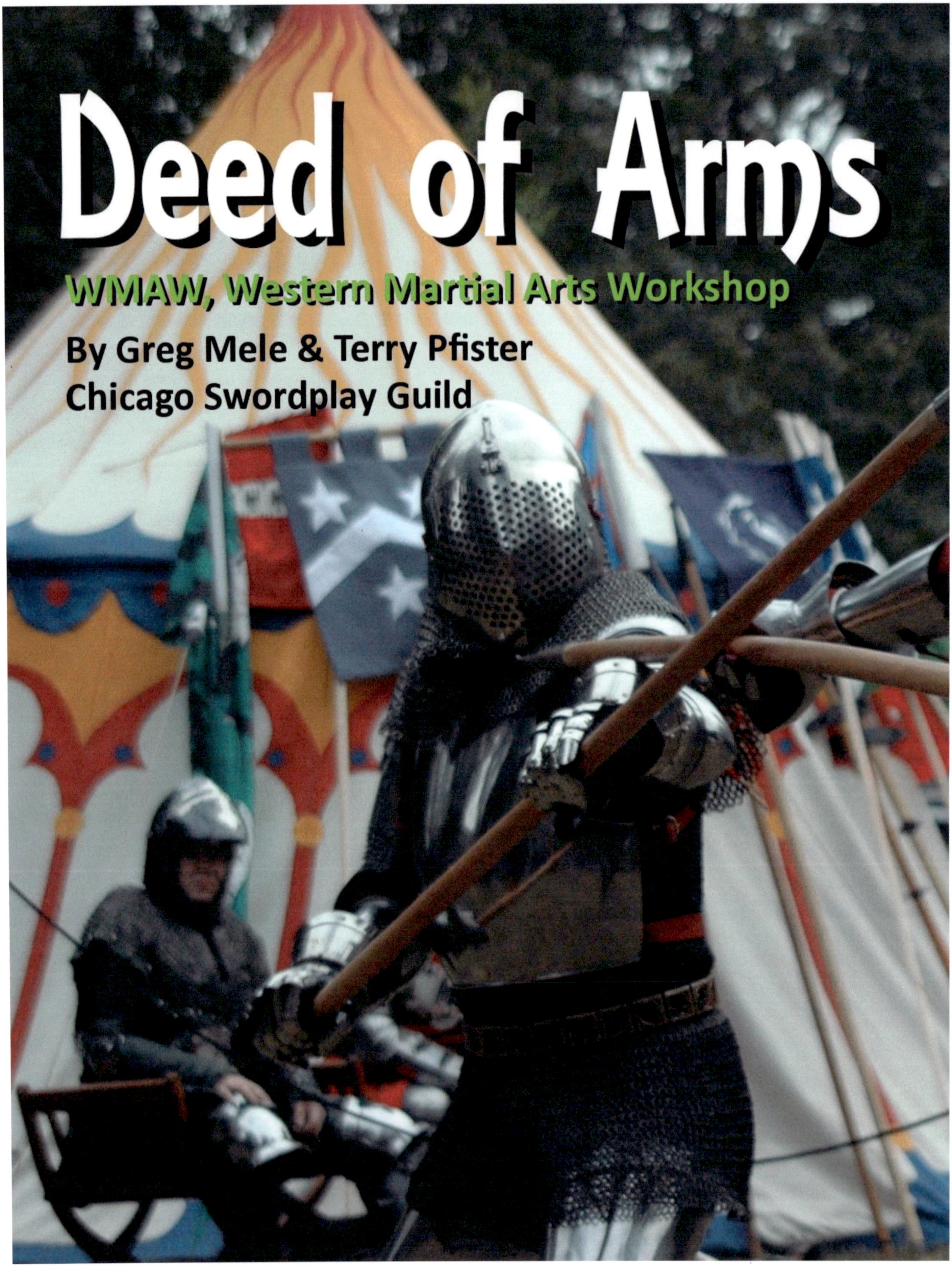

Photos Courtesy of Terry Pfister
Communications Officer
Chicago Swordplay Guild

Deed of Arms Writ Courtesy of Greg Mele
Co-Founder
Chicago Swordplay Guild

Sean Hayes (left) and Matthew Cacy (right)

The knightly tournament originated as mounted melees fought to capture opponents for ransom, and often became little more than an (ostensibly) agonistic battle fought for fun, fame and profit. By the 14th century, the role of the tournament had changed becoming less training for war and more of an engine for renown by individual and small groups of knights. More and more regulation was found, and there was more emphasis on individual combats.

From the 14th century through the third quarter of the 15th, a specialized kind of deed of arms became popular, one that strove to bring out the breadth of chivalric virtues within the knights, to showcase individual prowess, and to evoke something of the 'ideal' chivalry buried in the legends of history.

> Accounts of knights meeting head-on in a clash of splintering lance and cleft shields is a powerful icon of the knightly culture.

The pas d'armes ("passage at arms") probably grew out of the knightly tradition of single challenge. Early accounts of knights meeting one another head-on in a clash of splintering lance and cleft shields held a powerful place in the iconography of the knightly culture as it rose and changed from the 9th century to the 16th.

One or more defenders, known as the tenans, would make it known that they would hold a particular place on a particular day against all challengers, or venans. Such challenges were designed to celebrate the office of knighthood through a demonstration of arms. This was no game. It was more structured than the old tournament of the 12th and 13th century, and like the joust, the focus of the passage at arms was to showcase the martial prowess of the individual combatants. It was also a chance for those combatants to display

Bill Grandy (left) and Sean Hayes (right)

their courtesy, honor and largesse, which were the critical qualities that set a man apart from a brute.

To host a pas d'armes, men would first declare their intention, either at an event such as a feast or a knighting ceremony, and then set about spreading word such that other knights and squires could attend.

They might set up a simple pavilion at a crossroads, or they might sponsor a large festival of arms, which might last as long as a month. They would then arrange for shields, spears, weapons and sometimes armour to be available for their erstwhile challengers. Refreshments and gifts of extreme generosity were often given-it was an opportunity to display hospitality and the virtue of largesse.

Holding themselves ready for the hour, day or even the month of the challenge, the 'tenans' would await their opponents, who were met with friendship. Both prowess -skill in arms- and courtesy were important.

There are many accounts surveying the ferocity of the combatants, but side-by-side with these accounts is equal space devoted to the largesse and courtesy of even the most deadly combatant.

And, although there was often a 'gallery' of spectators present, as often as not there were no spectators-it was held for the joi de combat and for the glory and renown to be earned in such honorable combat.

Besides being a tribute to our ancestors, we also believe that this format is a useful tool for combatants to use the deed of arms as a test of not only prowess, but character. In the pas d'armes, combatants challenge one another for the pleasure of the combat-not for victory. In reality, other than the winner himself, few remember who one any given tournament, but all remember an exciting, invigorating – or embarrassing – bout, and the names attached to that bout. Renown is earned within the lists by demonstrating great skill (prowess) combined with the other chivalric virtues of, courage, generosity, humility (in accepting defeat or victory), faith

(in our ideals), a sense of justice and the duty of defense. These are the real contests of the tournament, and it is the yardstick by which we are measured.

The 'gallery' and your opponent can see who you are; it is hard to be deceptive of your motives and your sincerity in such contests.

WMAW armored combat is *not* a reenactment or living history presentation, it is more about practicing a martial art with an understanding of its history and context, rather than recreating historical times . Bringing these ideas from history and grounding them in the present is one of the best ways we can honor the arts we practice. To this end, the format, armor, banners and attire of the support staff (the heralds, valets to help people arm, judge, etc.) were historically based, that of the spectators and those off of the field was not.

WMAW armored combat is about practicing a martial art with an understanding of its history and context, rather than recreating historical times.

As a comparison, when a koryu demonstrates kenjutsu, they are wearing hakama or, with an art like Yagyu Shingan Ryu, real yoroi, but they are not pretending to be long dead samurai - they are doing their Art's history homage.

There were two teams, one of defenders and one of challengers. The nine challengers each got to challenge one of the seven defenders to combat with the longsword, spear, pole-

Devon Boorman (left) Leo Lastre (right)

Bill Grandy (left) and Sean Hayes (right)

axe or dagger. Heralds (one for each team) announced the fight, and valets brought forth the two combatants banners, which were planted at the edge of the field. Once the fight was over, the herald for the victorious combatant announced the victory.

The combat was based on conventions used in 14th and 15th century deeds of arms. What this means is that "victory" could be determined as follows:

1. Win the majority of five good blows landed with the point of the spear, dagger, sword or axe in the gaps in the plate armour, or to the visor of the helmet, or landed with the hammer head of the poleaxe *anywhere.* Cuts with the sword are ineffective against plate armour, so cuts only counted against areas protected by mail.

2. Disarm the opponent. Note that if he was still wearing a dagger, he could continue combat.

3. Drive the opponent out of the lists - a term called "being given the gate"

4. Throw the opponent to the ground and either remain standing, or land in a position of control. There was no ground fighting permitted.

The opponent could yield for any reason, at any time.

5. All of these victory conditions came into play over the course of the day.

The judge called blows, *but* a combatant could overrule him to his own disadvantage. What this means is that I could call a blow against myself that the judge did not see, or I could negate one scored against my opponent that I thought was not good. The purpose was to keep the combatants, and their own honor, a valid part of the deed of arms, and not make it just about playing the points.

Finally, unlike a modern tournament, there was no winner, per se, but rather the challengers chose to call out and acknowledge someone whom they thought fought best.

The challenge was a way to celebrate one's knighthood through a demonstration of arms.

Devon Boorman left, Nicole Allen center, and Matt Cacy right

International Battodo Federation
Toyama-ryu & Nakamura-ryu Taikai
2012 Washington D.C.

剣魂

NORTHERN VIRGINIA KENSHINKAI

International Battoudo Federation

Capital Area Budokai 武道

活人剣

Saturday, March 31, 2012 · 8:30am - 5:00pm
Hilton Alexandria Mark Center, Alexandria, VA
International Batto-Do Federation (IBF) Nakamura-Ryu Taikai

INTERNATIONAL BATTO-DO FEDERATION NAKAMURA RYU TAI KAI
Washington D.C.—31 March 2012

2012 will be an important year for the International Batto-Do Federation. The IBF is the parent organization of Nakamura Ryu Batto Do, based out of Yokohama, Japan. 2012 will be the 100th anniversary of the birth of its founder, Nakamura Taizaburo sensei.

In addition to the annual Tai Kai (competition) in Yokohama in October of next year, there will also be a very special event in the U.S.

The Virginia Kenshinkai, a full member dojo of the IBF, is pleased to announce a U.S. IBF Nakamura Ryu Tai Kai, to be held in Alexandria, VA (near Washington D.C.) on March 31st, 2012 (registration and lodging information available soon). The unique aspect of this tai kai is that it will be strictly according to the rules and guidelines of the IBF. That means required individual forms at each level, required individual cuts, one competitor at a time, graded by a panel of Japanese sensei. Plus team shizan and kumitachi. Details will be posted on our website Capital Area Budokai and the VA Kenshinkai.

We are expecting a large contingent of senior Japanese sensei to attend, judge, teach, and participate,

DAVID DRAWDY (LEFT) AND CARL JENKINS (RIGHT) COMPETE IN KUMITACHI

as well as many of our international members and our friends and fellow kenshi from around the US.

We have permission to open this event to non-IBF members and will be offering seminars on the required tai kai forms and events in various locations around the US prior to the event, as well as seminars taught by the Japanese sensei immediately before the event. In addition, we are inviting a broad spectrum of groups, artists and craftsmen from around the US.

FRED MCLAREN LEFT AND TOM WADLOW RIGHT COMPETE IN KUMITACHI

Anyone interested in attending or participating, look for information or our website Capital Area Budokai and the VA Kenshinkai or feel free to contact me via PM on the Japanese Sword Arts section of SwordForum International, or Carl Jenkins-sensei, senior instructor of the Kenshinkai, who is doing most of the planning for this event.

Please note we are a registered non-profit organization and will gratefully accept your tax-free charitable contributions.

Thank you,
David Drawdy Sensei

David Drawdy Sensei titles include:

⇒ Sergeant Major, U.S. Army (Ret.)
⇒ Godan Renshi, Nakamura Ryu Batto Do
⇒ Shibu-cho, International Batto Do Federation
⇒ Dojo-cho, Virginia Kenshinkai
⇒ Director, Capital Area Budokai

www.capitalareabudokai.org

CARL JENKINS SENSEI THOMAS WADLOW DAVE DRAWDY SENSEI

Creative Commons Credit Michael Shanks

Excalibur

" Whoso Pulleth Out This Sword From the Anvil That Same is the Rightful King of England"

History of the Japanese Blade

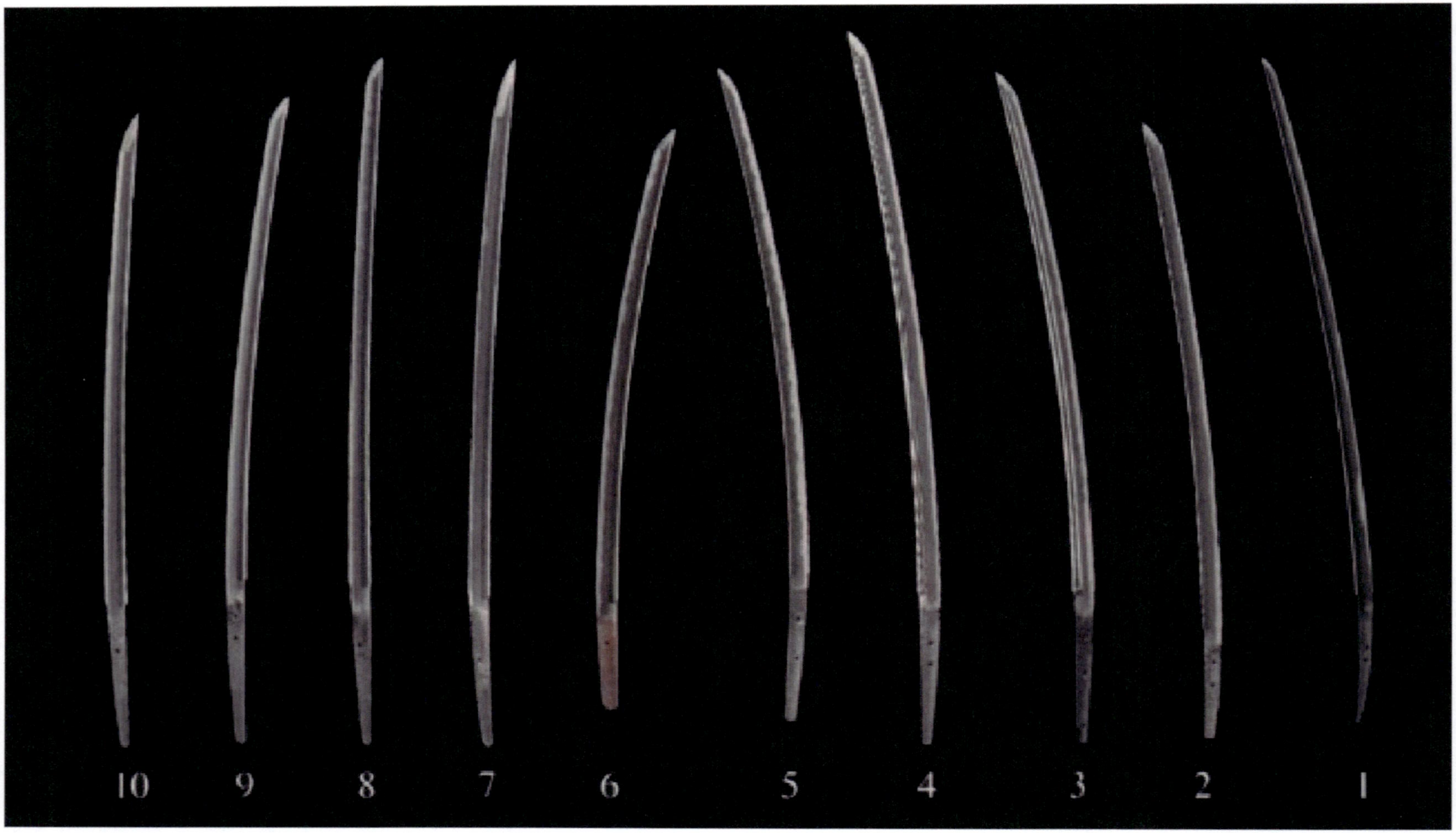

Katsujinken, a Sword Arts Journal presents. . .

Paul Martin, Japanese Sword Curator | Researchist | Former Japanese Antiquity Member (British Museum London England)

AN OVERVIEW OF JAPANESE SWORDS

PAUL MARTIN
Photo: Tom Kishida

0. JÕKOTÕ (NO IMAGE)

Jõkotõ are straight blades with no curvature. They are usually constructed in the hira-zukuri and kiriha-zukuri styles. It is thought that the shift from straight blades to Japanese swords with curvature happened around the mid to late Heian period (794-1184). This was during the mid 10th century: about the time Taira Masakado and Fujiwara Sumitomo rebelled against the government in the Johei (931-938) and Tengyo (938-947) eras. Blades before these are continental style blades called jõkutõ, and were brought to Japan from the Asian continent. Examples of these blades have been excavated from Kofun period tombs, and some still reside in the Shosoin Imperial Repository, Nara.

1. LATE HEIAN TO EARLY KAMAKURA

From the late Heian period and the early Kamakura period (1185-1333) we can see the Japanese sword as we know it: shinogi-zukuri (ridgeline) construction, with a wide base, narrowing acutely towards the small point section (ko-kissaki). They are quite slender blades with the curvature concentrated between the handle and base. This shape is called koshi-zori. From midway towards the point there is very little curvature. These blades are usually around 2.5-6 shaku in length (75.8 cm-78.8 cm).

2. MID-KAMAKURA

At the zenith of the warrior class's power during the Kamakura period, the blade's kasane becomes thick, the mihaba becomes wide and they take on magnificent tachi shape. There is not much difference between the size of the moto-haba and the saki-haba. The blade still has koshi-zori, but the center of the curvature has moved further along the blade. The kissaki has become a compact chu-kissaki (ikubi). The hamon has developed into a flowing gorgeous choji-midare. Also around this time, tanto production appears.

3. LATE KAMAKURA

Tachi at the end of the Kamakura period have developed into magnificent blades. There are two types: one is wide throughout its length and the point section is the same as mid- Kamakura period kissaki, but slightly extended. The other is quite slender and similar in appearance to the late Heian, early Kamakura shapes. However, when you look further along the blade the shape has changed; the curvature has moved further along the blade. During this period notare-gunome hamon appeared. It is said that in Sagami province Goro Nyudo Masamune perfected the production of nie-deki blades.

4. Nanbokucho

During the Nanbokucho period many long blades of three shaku (90.9cm) and other long tachi were made. Tanto of large proportions were also produced. Tachi were majestic, wide, and proportionally long. Among these were some over 90 cm in length and worn over the back. These types of blade are called no-dachi and o-dachi. They were rather thin in construction to decrease the weight. Additionally, many have a bo-hi (groove) cut into the shinogi-ji area in order to lighten the blade. Many tachi from this period are o-suriage (shortened in later periods as they were difficult to wield). Consequently, many extant blades from the Nanbokucho period are unsigned.

5. Early Muromachi

Blades of the early part of the Muromachi period are reminiscent in construction to the blades of the early Kamakura period. When compared to the shape of the Nanbokucho period, the shape has completely changed and no longer includes o-kissaki. At around 2.4-5 shaku (72.7 cm-75.7 cm) in length, they are quite narrow and deeply curved with a medium-sized point section. At first glance they may appear somewhat similar to Kamakura period blades, but on closer inspection they are saki-zori character.

6. Late Muromachi

By the late Muromachi period, samurai fighting methods had changed from cavalry to mass infantry style warfare. Uchigatana, worn with the cutting edge uppermost thrust through the sash, had become popular. After the Onin war, conflicts broke out in many places and kazuuchi mono began to appear (mass-produced blades inferior in quality to regular Japanese blades). However, specially ordered blades of excellent quality (chumonuchi) were also produced at this time. Bizen (Okayama prefecture) and Mino (Gifu prefecture) were the major places of production. Many blades produced in this period are around 2.1 shaku (63.6 cm) in length. They are slightly wider than the standard width, with either a chu-kissaki or an extended chu-kissaki and strong saki-zori. The nakago are short, intended for one-handed use.

7. Azuchi-Momoyama

Swords produced up to the Keicho era (1596-1614) are classified as kotõ (old blades) Blades made during and after this era are classified as shintõ (new-swords). When Japan entered the Azuchi-Momoyama period, many smiths moved to Edo or Kyoto, or gathered in castle towns of various influential daimyo. Additionally, developments in transportation brought about experimentation with materials, and foreign-made steel (known as nanbantetsu) was utilized. The blade's shape from around this period mirrors that of shortened Nanbokucho blades. They are generally wide with little or no difference between the moto and saki-haba. Many have an extended chu-kissaki, whilst some have o-kissaki, with a thick kasane and are usually around 2.4-5 shaku (72.7 cm -75.8 cm) in length.

8. Mid-Edo

Swords of the mid-Edo period are of standard width. The saki-haba is relatively narrow when compared to the moto-haba. The curvature is noticeably shallow with a small to medium-sized point section. They are usually around 2.3 shaku (69.7 cm) in length. This particular type of construction was generally produced around the middle of the Kanbun (1661-1673) and Enpo (1673-1681) eras, and is usually referred to as Kanbun shintō.

9. Edo Period Genroku era

The change in shape of Japanese swords between the Jokyo (1684-1688) and Genroku (1688-1704) eras reflects the transition of shape from Kanbun-shintō blades to the beginning of the shin-shintō period of sword manufacture. As it was a very peaceful period in Japanese history, rather flamboyant hamon appear, and as opposed to that of Kanbun-shintō blades, the curvature is quite deep.

10. Edo period Bakamatsu era

Blades made after the Bunka (1804-1818) and Bunsei (1818-1830) eras are referred to as fukko-shintō (revival swords). Pioneers of the revival movement include Suishinshi Masahide and Nankai Taro Tomotaka. Taikei Naotane was among Masahide's students. Minamoto Kiyomaro led a revival aimed at soshu-den and Mino-Shizu workmanship. Bakumatsu blades are shallow in curvature, have a wide haba with not much difference in width between the saki and moto-haba, and are around 2.5-6 shaku (75.7cm-78.7 cm) in length, with an o-kissaki and thick kasane.

11. Meiji Onwards (no image)

Blades made from the 9th year of Meiji until present day are referred to as gendaitō (modern swords). As of the Hatorei decree in 1876 (banning civilians from wearing swords), the need for swords declined. However, in Meiji 39 (1906) the craft gained imperial patronage. The Swordsmiths Gassan Sadakazu and Miyamoto Kanenori were appointed Tei Shitsu Gi Gei In (craftsmen by imperial appointment—equivalent to National Living Treasure). Since then, the swordsmith's craft has continued through the Meiji (1868-1912), Taisho (1912-1926), Showa (1926-1989) and Heisei (1989-) eras until today. Today's Swordsmiths try to recreate the workmanship of eminent smiths of every period, regardless of whether they are kotō, shintō or shin-shintō. In particular, recreations of tachi of the Kamakura period are a popular aim for many modern Swordsmiths.

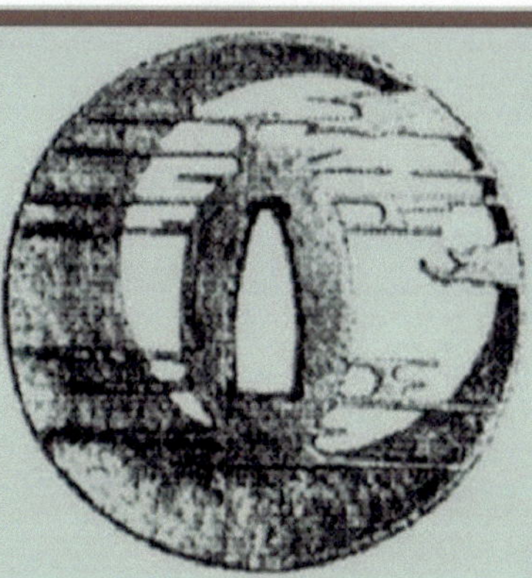

KATSUJINKEN SENSEI SPOTLIGHT

BOB ELDER SENSEI

Katsujinken—*a Sword Arts Journal* interviews an Instructor or Sensei from either the Eastern or Western Mixed Edged Weaponry—Sword Art practices each quarter. For this, our "Premiere Edition," Katsujinken has put our Katsujinken Spotlight on Bob Elder Sensei of Orlando Toyama Ryu Dojo, Orlando Florida.

We interviewed Elder Sensei during the summer of 2011. Elder Sensei, a veteran Martial Arts and Swordsman—Bob began studying as early as 1972. When many of today's practitioners were a concept—Elder Sensei was already hard at work. Katsujinken is proud to present Elder Sensei—a seasoned expert—as he shares true wisdom and knowledge from a lifetime of training, study, and teaching. His unabashed communicative style is well-known and respected . . . Elder Sensei presents straight from the hip facts and L2 (lessons-learned) in this insightful reflection of his Swordsmanship.

1 How long have you been a practitioner of the martial arts?

In 1972 I began 'real-training' in the martial arts. So . . . Jason, Katsujinken can safely say I have been practicing martial arts for thirty-nine years. And yes, this is a significant amount of time although I really don't view in those terms.

2 What first got you interested in swordsmanship and what has kept your interest going so strong through-out the years?

Initially, I was offered to a job delivering school supplies as a fill-in for a friend of mine (he'd recently wrecked his Harley Davidson). On a delivery, I noticed a two by four sticking out of the ground at one of the warehouses. It seemed odd to me and I began inquiring with the boss who was also the business-owner. He ignored me for the majority of the day, but I wasn't taking no for an answer. Finally, he told me I had bugged him enough and explained what the setup was. This is where my knowledge of Isshin Ryu Karate began and my first introduction to Bob Earls as an expert in the art was founded.

BOB ELDER SENSEI

Photo Courtesy of Doug Waller

From this initial introduction - I finagled my way into classes with him at his house. And, I'm glad I continued to pressure him at that time, because a short-time later, he began a dojo and went public with training. Eventually, I was introduced to Bob Earls teacher—Bob Bremer of Miami.

Bremer was the real deal. I was fortunate to begin training with him when he moved to Gainesville and then later in Ocala Florida. Bremer features a really cool, unique, and distinct history. He lived in Okinawa for seventeen months as a civilian and trained with Tatsuo Shimabukuro. He also taught classes in Okinawa – which is incredible.

Green belt in Isshin Ryu karate. 3rd place trophy. Circa 1973

Bremer got"Karate – he was mind-blowing. He had learned 'chinkuchi' a\s an Okinawan, very different from an externally trained master. His understanding of physicality of Karate applied to all Japanese and Okinawan Martial Arts, though his discipline is primarily Okinawan. I found, related and continued to teach this same ideal of movement in my classes, because it works.

Over the years, I have not found anyone in the Japanese Arts or any other of the physical foundations perform as correctly. Maybe they all do, and I am just unaware. Martial Arts are my life and I know many different people, which has allowed me an opportunity to get 'in' with many of trainers without being a direct student. Through these allowances . . . I have never been made aware of anyone with a more significant in-depth (spiritualesque) understanding of what we (they) are doing.

To me, the goal is to discover the 'axis' of Martial Arts and how they speak as an Artform—with underlying spirituality as thoughts of Zen or Tai Chi. All of us practice a 'spoke' of truth when practicing our varied Art Forms. Rarely does anyone 'get to the axis' or 'crux of real truth.' In our studies (as an industry), 'they' (meaning mainly instructors) tend to teach and propagate the nature of the Artform most familiar or where the most buy-in is maintained. Sometimes, teachers each do the same things, as the consensus is to instruct what is commonly accepted and not deviate from the norm. So each generation of instructors tend to bleed out the realism of Sword Arts. And, every subsequent generation of instruction, I have learned in my thirty-nine years of participation, tend to water down real knowledge for 'expected knowledge.' The reasons for this I affirm are not necessarily intentional—it is due to many reasons . . . such as not understanding what they see in training and then instructing a misappropriation of the definition of the Art or they lack real depth in the Artform.

Where depth and understanding dissipate in Martial and Sword Art studies, the concepts in multifaceted comprehension are lessened. This results in a loss of the overall Artform dynamic. The 'axis or core of studies,' again become lost. Then a homogenized replacement delivers a vanilla regimen in training. No one is at fault for this reality, all of us in the industry are guilty of a sort of Dumbing down of the Artforms . . . our students and the future of Martial and Sword Arts suffer as a result. There is a real sin being committed when students do not earn a true understanding of the Martial and Sword Arts.

3

So, what first got you interested in the Sword itself, how did you meet Hataya Sensei for the first time?

My interests in Japanese Sword Arts stretched over a 15-year span where I collected and studied Japanese Antiquities. In a chance commentary from a Nisei friend of mine during a sword show in Houston, I became introduced to Hataya Mitsuo (a 6th degree sword master). This introduction shook my world. Truly, a spooky shiver (and all that goes with that) went down my spine. During the sword show, I learned Hataya would be in Tampa at another sword show, so I invited him to Orlando to train. He agreed.

And, this was a beginning of a dream come true for me and others wanting to learn Japanese Sword Arts. His agreement to come to our neck of the woods caught me off guard, so much so, I almost tossed my cookies.

While I respected the Karate Instructors, I knew, I had decided I did not want to learn the Japanese Sword Arts from my friends who only sidelined training in Japanese Sword Arts. So, when Hataya Sensei came to Orlando the next February – I arranged a seminar. Right off the bat, thirty students showed up to train, so we hit the bricks running. Just this year, I was promoted to 7th Dan . . . so the early introduction to Hataya Mitsuo to come to Orlando was fortuitous and I believe providence reigned.

Photo courtesy of Víctor Herrero Pérez Sensei
Toyama Ryu, Spain

4

You have your own dojo? What's the name of it? How might our readers locate it if they were in the area?

My Dojo is in Orlando. It conjoins my company East Coast Martial Art Supplies located at 1646 East Colonial Drive, Orlando, Florida (USA) 32803. Everyone is welcome to reach out to us at (407) 896-2487 for more information. Our website, currently under development, is www.orlando-toyamaryu.com.

5

What do you think about the students coming up today (the younger generation)? We're sure you have had many students who have made you proud over the years. Certainly, at this year's past Orlando Tai Kai we noticed many very talented Toyama Ryu | Batto Do practitioners; some of whom came up under your tutelage—how does this make you feel?

Earlier in the interview, I spoke a little about getting to the 'core or axis' of training and that applies here; however, I feel students learn faster now with training being offered in English. Bob Lampp and I started together in 1991. We had to learn the Japanese Sword Arts by observation alone. We often stumbled through misguided . . . non-budo understandings via Japanese students as translators.

At my dojo, I want students to gain depth and not just garner a glossy façade while training. Still, anyone born after 1990, is raised on technology. Most of their day is spent on a screen of some type—this makes their transition to Martial and Sword Arts somewhat difficult in general.

At Orlando Toyama Ryu Dojo, we have ways to aid our students out of the screen or technology mindset—to the Budo mindset.

6

We know you own and operate East Coast Martial Arts - how has business been for you being involved in the Martial Arts industry?

ECMAS (East Coast Martial Arts Studio) began in 1979, when I started it as a business. Due to this, I have had the ability to make Martial and Sword Arts my life. This is not a hobby to me and I do believe I approach Martial and Sword Arts in a way only a few people comprehend. I have had the good fortune to get to know many Martial and Sword Art practitioners without being in their dojo, kind of like a fly on the wall. I've gotten to participate in many great conversations over the years with industry leaders while securing training insights others might not gain exposure to. ECMAS has presented an ideal situation to me and added a depth to my understanding I could not have bought nor afforded otherwise.

7

What could you tell our readers about the USBDF (U.S. Batto Do Federation) and how it relates to some of the other Sword Arts? You were one of the founders and long time president, is that correct? How has it grown and where do you see it leading as an organization in the future?

I founded U.S. Batto Jutsu Federation and modeled it after Zen Nihon Batto Do Renmei of Japan in order to make Tai Kai attendance easier. It is "only" a technical association and does not teach the 'heart, axis or core' of the Japanese Sword Arts.

For instance, if you are a Nakamura Ryu MJER or any 'other' sword style – you learn kata and cutting patterns through USFBD in order to compete with one another among competitors each doing the same thing. This allows judges to judge better. Without the uniformity, judges would find it difficult to judge one Sword Art style from another.

For example, is your foot in or out, sword held correctly or incorrectly? How is your grip? Without the USFBD as a technical association—one grip in one dojo wouldn't map to another grip in another dojo, but neither would be incorrect . . . so, the USFBD aids Japanese Sword Arts through a semblance of order, which provides the base for judging and competition. However, this does not necessarily equate to the 'true axis or heart of training' as individuals gaining their relationship to their Martial and Sword Art forms—as practitioner.

USFBD does present an event where many people from many differing styles are able to come together and learn. The camaraderie is exceptional, it is a great place to be for a Japanese Sword Art practitioner.

To make USFBD more successful—at times I had to wrestle personalities and individuals with 'titles' to keep the association 'real.' However, since it is not my brand—and reads as Hataya Ryu—in the Budo . . . heart of training—I work to stay out of the way of personalities. Just as in any other sport affiliation—personalities creep in at times to create havoc and ignore the overall good of the Art itself.

8

Finally, Any thoughts, philosophies or ideas you would like to share?

Today, I find new students are usually young and any Samurai experience they have is often Anime. To me, the way a person gets started in Martial and Sword Arts—the initial reason for their motivation—is okay if their heart is okay. The real duty and 'Giri,' an instructor needs to pass to students—is their years of experience through an unedited heart and soul in training.

What is the reason?

Why bother?

Swordsmanship is not for self defense.

There is no money in it, no chicks and no glory. Well, some trainers derive glory where they themselves have not had solid growth and I have to be honest—there are many of those around. Usually, these trainers are addressed as 'Kyoshi,' or 'Master' as if that is their name. But, in Japan, this is not done. You may ask, Why? The reason is—a small minded person can be a big shot to the uninitiated—this is a sad reality and one that occurs in all walks of life. Look at the religious zealots who have duped good people through personal vainglory.

Nevertheless, those with heart, the 'real deals' have to stay true to the way of their Art Forms. To do so, is to not partake in arguments garnered by those seeking vainglory.

So . . . what are we doing overall? We are keeping an antiquated Art Form alive.

In doing so, we shouldn't attempt to alter the Art Forms—instead, we should wholeheartedly pursue their true understanding, respect and conveyance of knowledge. We need to represent Sword Arts from our hearts, while recognizing the 'axis or core' of their composition.

We need to strip away negative personality traits such as willfulness or wantonness. We need to ensure we use positions respectably among the Sword Arts community members. There is no justification in belittling lesser ranked students.

And, there are absolutely 'no excuses' for breaking one of three cardinal rules when serving as a leader in the Mixed Martial and Sword Art community; these are: 1) sex with students—every Instructor and Sensei worth their rank find enamored students—breaking this cardinal rule is where the proverbial excrement hits the fan; 2) using position of Instructor or Sensei to gain inappropriate position over students (e.g., allowing wealthier students to buy things for the Dojo or Instructor and giving special treatment for their doing so—breaking this cardinal rule results in a lack of overall Dojo respect; and 3) letting cliques develop—to break this cardinal rule is just bad business all the way around.

This is just Bob Elder Sensei take on things . . . I come at this with literally decades of study. A student has a duty as a student and an Instructor has their duty also. The lines of consequence cannot cross. Martial and Sword Arts are excellent additions to anyone's life who garners a heartfelt desire to participate. For most part—the heart puts them in the right place at the right time with the right results.

Bob Elder Sensei

Bob Elder Sensei, *Katsujinken, a Sword Arts Journal* thanks you for agreeing to participate in our Premiere Sensei Spotlight—we appreciate the magnitude your career has traversed in both Martial and Japanese Sword Arts. We are truly honored for the time you have provided to us and for all of the dedication you have demonstrated to the arts.

Bob Elder Sensei has been more than just a Sensei to me in my life. As much as he has taught me about Swordsmanship and he has given me many lessons outside of the life of swinging a sword.

Since the first time I visited his shop in the late 90's, he has taught me how to enjoy life instead of being just so caught up in it that you lose focus by focusing too much . . . such as, how to drink beer, what qualities it takes to be a good teacher, and how to appreciate time as a student. He also has taught me, being able to laugh at yourself is as important as being able to make others laugh. Most of all, he taught me how much a gift it is to practice good budo while not becoming a "just" a great practitioner, but also to be a qualitative person people rely upon.

~Sang Kim, Byakkokan Dojo, New York City, New York.

全米抜刀道連盟

www.usbattodo.com

Seishinkan Dojo
Bob Elder Sensei
Orlando, Florida

Byakkokan Dojo
Sang Kim Sensei
New York, New York

Bamboo Dojo
John Matthews Sensei
Vero Beach, Florida

Genbukan Dojo
Douglas Firestone Sensei
White Plains, New York

Seibukan Dojo
Tom Smyth Sensei
Palm Bay, Florida

Bluff City Iai Kai
Mark Chevalier Sensei
Memphis, Tennesse

Seiryuukan Dojo
Robert Steele Sensei
St. Paul Minnesota

Konjaku Kioi Dojo
Mike Femal Sensei
St. Petersburg, Florida

Melbourne Dojo
Lois Fredricks Sensei
Melbourne, Florida

柄

Making a Tsuka

by Christopher Treichel

Christopher Treichel is an accomplished battodo swordsman and Revolutionary War Re-enactor. Chris developed his woodworking, leather working and related skills by making arms, equipment and clothing for his reenactment interests. Over time his craft naturally found an additional outlet in Japanese sword related items as he developed in his practice of the Japanese sword arts. Chris has done numerous sword mountings, tsuka, hakama, and sword belts. Chris is truly a renaissance man.

Much like many Japanese arts this is not something really difficult but can be very complex the longer you think about it. Everything involved in making the tsuka is there for a purpose… one item supporting the next to allow the user to safely wield a sword. I am but a student and here is how I do this. Perhaps someday I will learn to do it right.

The wood I use is poplar also known as tulip wood as honoki is hard to obtain in quantity and quality. Poplar has nice consistent grain, is dry sap free wood and carves well. When selecting the wood it is important to find lumber that has straight grain, with no knots and no run out (where the grain cuts across the wood and will split if subjected to force). Once obtained allow the wood to sit in the environment where it will be used or stored for several months so that it acquires the humidity and temperature of that location. The carving needs to be done very exactly, in essence you are inletting the nakago (tang) into the tsuka and the fit should be tight enough that the yarisuri mei (file marks) and pressure of the wood will hold the nakago in place and not rely solely on the mekugi (bamboo pegs).

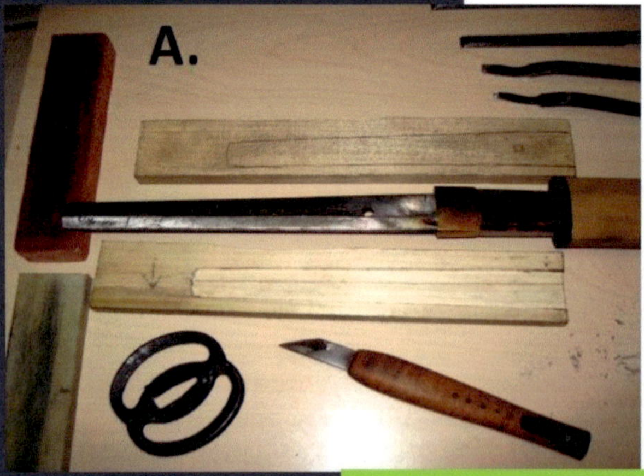

Before starting to carve you need to think about the type of tsuka you will carve… will the sides be parallel, both curved or only one curved? How do you want the nakago to align inside the tsuka? What type of fittings will you use? Will you utilize this sword for decoration or will it be used for martial arts? Will you apply a full same gawa wrap or only panels? You start to get the picture that you can't just do a lay out and start hacking away until it looks right.

In this case the Ha (sharp edge) side of the tsuka will be straight and the mune (back) side will be curved. I will be utilizing a full wrap for martial arts use. The Final wrap will be katatemaki (battle wrap) with silk ito.

Inletting the Nakago

Pic A: You will need a few tools… bent chisels and paring chisels as well as a carving knife will be needed as well as a sharp pencil. Due to not wanting to scratch up the blade with sand paper particles you will also need a furniture scraper which is nothing more than a piece of spring steel that has sharp sides.

Pic B: I trace the outline of the nakago onto one half of the wood for the tsuka using the carving knife so as to be exact. I then go back with a pencil to darken this line. I add a line for the shinogi (rigeline) to allow me to see where I need to carve deepest.

Pic C: I use a small bent paring chisel to cut down to slightly more than half the width of the nakago in the area between the mune side and the line I drew for the shinogi. I use a larger paring chisel to remove the wood at an angle from the shinogi line up to the ha edge.

Repeat this for the other side but only cut slightly less than half the depth. This allows the edges to but up against a more solid side than if you had simply cut both sides equally to half the depth.

Pic D: you can see that I am planning where all of the parts will go. Here you see the outline of the fuchi traced onto the tsuka core

Pic E: You see that I am considering where the nakago is in relation to the outside of the tsuka core to drill the mekugi ana (peg hole). Once the mekugi ana are drilled it's time to glue the tsuka together. I prefer a water resistant wood glue like Elmers or Titebond outdoor wood glue. Clamp and let dry for 24 hours.

Carving the Tsuka

Pic F: First you will need to line everything up and inlet the kachira until the ana (hole) in the nakago and tsuka line up. Mark out where the hand placement will be to ensure that the curvature corresponds. Carve down to the ha and mune sides. It helps to work in two dimensions at a time so you can measure the diameter and won't accidentally carve down at an angle to the nakago.

Pic G: Carve down the sides of the tsuka. You will need to take into account if you are going to use panels or a whole samegawa wrap so that you carve down the tsuka to the proper diameter. A tool that will help you keep the sides even is a small hand or finishing plane.

Pic I: A good thing to do is to start planning out exactly where things are going to go... After remeasuring the length according to the width of the ito (string covering the tsuka) and the number of turns about the tsuka you can figure out where the kachira is to be and can now be inlet.

Samegawa

Pic J: After the kashira has been

B.

C.

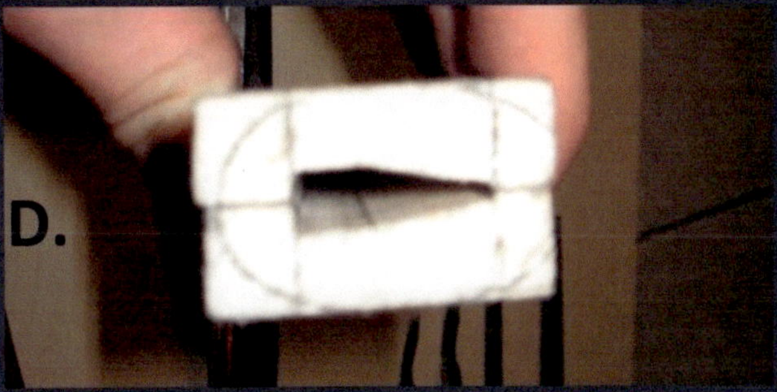
D.

E.

F.

inlet and the thickness of the ito and samegawa have been taken into account you can carve down the tsuka proportionally

Pic J: Apply the samegawa with the seam on the side facing inboard when the ha is up. I use twine to secure it in place while the glue dries.

Pic K: After the samegawa is almost dry trim it to fit snugly. Once the same gawa is applied, add a thin piece of paper to each side of the tsuka so that the ito does not snag on it. This will allow you to really pull the ito tight without the same gawa abrading it. Once the same gawa is completely dry chamfer the mekugi ana with a reamer.

Tying Tsuka Maki

A short note... I am not going to go into great deal on how to tie tsuka maki as there is a great article from which I have gleaned most of my knowledge. Please refer to Thomas Buck's article ***Tsuka-maki, The Art of Japanese Sword Restoration.*** *Colorado Token Kai Quarterly,* v1, n1, pp. 3-8. I have also learned a great deal from suggestions on Sword Forum International. These are a great place to begin learning if this interests you.

Let us continue. In this case as mentioned, I am doing a variety called katatemaki which is a bit trickier than some of the normal versions.

Here are some recommendations on tsuka maki.

Above everything else the ito needs to be **TIGHT.** If you can move any of the crossovers with your fingers you are not doing it right. No amount of glue or epoxy can do this without looking like you glued everything down. Putting epoxy or double sided tape on the sides of the tsuka also does nothing to help you keep the string tight. Use your hands a clamp and a tying stand to really get things where they need to be. Most of this will require lots of practice until you develop a way that works for you.

Pic K: To make positioning the menuki easier hold them in place with rubber bands.

Prepare hishigami (small pieces of paper that lock in the crossings of ito). You need to take into account the width of the ito and how high you want the knots to ride when making the hishigami. The thicker the hishigami the higher your crossovers. Make all (60-120) of your hishigami before you start working. I would recommend giving it a few practice tries before jumping right in to ensure the

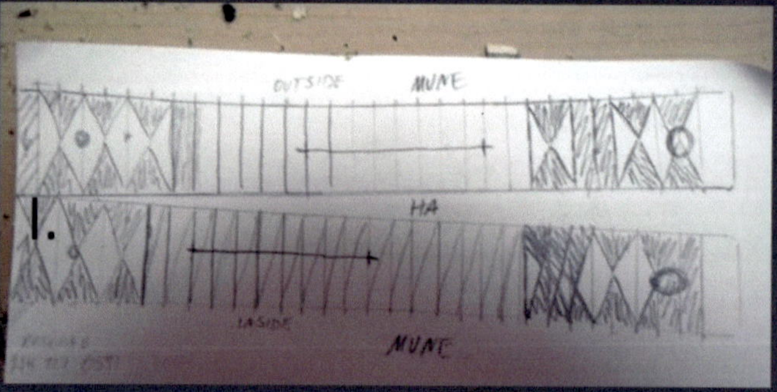

hishigame are the right size and support the crossovers the way you want them to. Hishigami are essential in maintaining the shape of the crossovers. If you apply them when they are damp they will lock the ito into place once dry.

Now you can begin tying the tsuka maki. This always starts on the side facing out when the ha is up.

Pic L: Ensure all the diamonds line up and are the same size. Ensure that the ito is straight on the sides of the tsuka

To make the ito stay in place when transitioning from two to one and one to two ends of ito doing a katatemaki style you add a dab of glue prior to securing the loose end.

Pic N: Once you complete the wrap you need to make any final adjustments to the kashira and cut the hole for the ito to pass through. If you are going to add shitodome to the kashira you will need to make sure they fit properly. Armorers files really help with this.

Some tips on tying the final Ura and Omote knots

Pic M: If you keep the knots loose until all the pieces of ito have been pulled through. You can then manipulate the ito so that everything falls into place... too tight and it will pull the knot up. Too loose and the knot will come undone. Here you will also need to support the knot with pieces of damp scrunched up paper almost like hishigami. If you do not use | insert paper into the knot it will come undone, slip or worse look ugly. Yet, once the paper dries it will keep the knot rigid.

Pic N: There are several methods for pulling the ito through the knots... I prefer to use a sail-makers needle and thread or twine. Others have luck using a bit of folded up plastic. You need to be aware that the same gawa is quite rough and will chew up your ito.

Pic O: Done

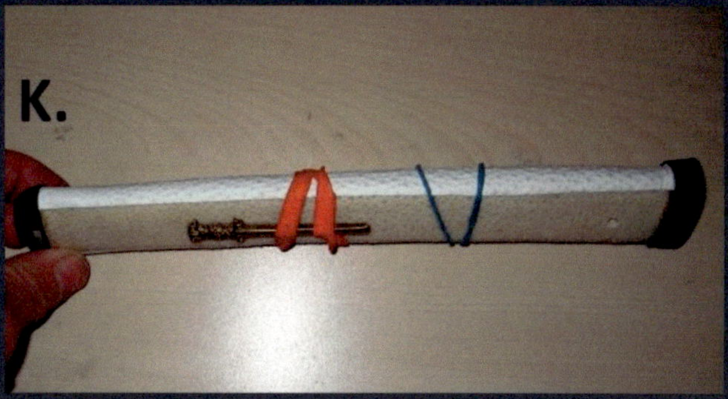
K.

L.

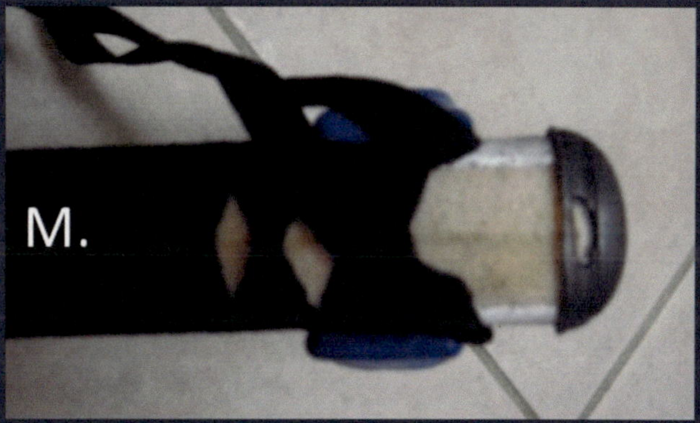
M.

N.

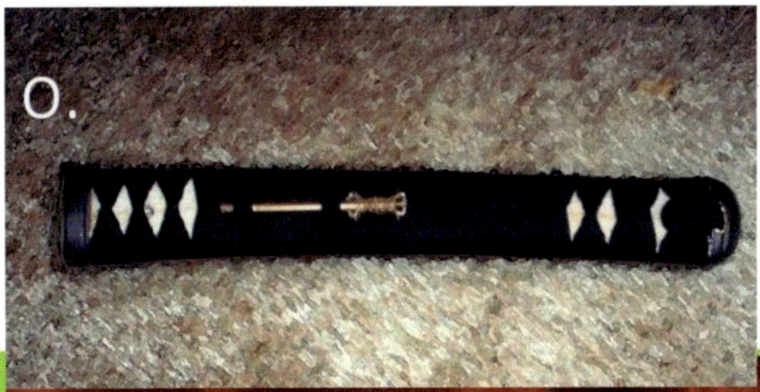
O.

無外

Mugai Ryu Gasshuku 2010

by Jason Lee A. Hatcher, Founder Katsujinken Magazine

The founder of Mugai Ryu, Tsuji Gettan was born in 1649 and studied Yamaguchi-ryū Swordsmanship under Yamaguchi Bokushinsai. At 26, he received Kaiden and opened a school in what is now Tokyo. He studied Zen and Classical Chinese under Zen monk Sekitan Ryouzen. At the age of 32, he reached enlightenment and received a poem from his Zen teacher. The name Mugai came from this poem.

Niina Gosoke of the Meishiha Mugai Ryu

Senpokan Dojo

The Senpokan Dojo was founded in 2003 by Tony Alvarez Sensei in Costa Mesa, CA. Prior to 2003, the Senpodoko Kai was in operation at the Shidare Yanagi Ryu Dojo of Don Angier Soke in Long Beach, California.

The name Senpokan translates as" to seize | strategy | school" "or in other words, "seizing the initiative through combative tactics and principles." It is from the term sen sen no sen that this philosophy of taking the initiative in battle or in life comes from and is a core principle behind all of the training and teaching within the Senpokan Dojo!

The arts taught and studied at the Senpokan are Meishi-Ha Mugai Ryu Iaihyodo, Toyama Ryu Batto Jutsu, the Seitei curriculum of the Zen Nihon Batto Do Renmei, Uchida Ryu Tanjo Jutsu and the Hojo Jutsu of the Shidare Yanagi Ryu Aiki Bugei.

Kneeling in seiza in the front of the Senpokan Dojo (Big Tony Alverez Sensei's Dojo) while in Costa Mesa, California was none other than Niina Gosoke of the Meishiha Mugai Ryu.

Niina Gosoke head of the Mugai Ryu Meishiha, wore a brown kimono and hakama with a personal crest. He carried a black folding fan. His daisho had gold themed hints throughout—complementing his overall appearance.

Off to his side, was Soke's right hand man—Renfield Kuroda of Japan. And, Bill Henderson Sensei, a native instructor from California as well.

Towards the back of the Dojo—lined in ranks were three rows of approximately 25-30 Mugai students. We all sat knees and tsuba's aligned—staged perfectly—listening for instructions. It appeared as if we had stepped into a feudal Samurai scene from a movie. Reminiscent of a Samurai clan awaiting orders from the Daimyo.

Each of us had traveled from distant lands to attend—some from Chicago, others from Wisconsin, two of our groups from Virginia attended, there were participants from Utah and California itself as well as guests from Japan.

Training began on a Tuesday evening at around 6:00 pm. The weather was cool with a slight breeze. There were absolutely no clouds in the sky. We left the Dojo doors open since the cool air delivered a needed break from all the sweating to take place indoors—while in training.

We began Gasshuku with kumitachi and kenjutsu. This provided all the wood-on-wood bashing we could ask for in any given evening. It was nice to get the fun stuff out of the way in order to lessen our anxieties from traveling and new introductions—this provided an outlet and ability to save the day, which would follow for the more dynamic training in seiza and tachi waza.

Renfield Sensei provided most of the teaching during our two days of training. Highly insightful, he regarded every movement with a watchful eye and resulted in our working together in a classic manner.

He also demonstrated how some kata, do not always work on a realistic scale until you apply the slightest of adjustments .

He explained, and then demonstrated influences he had received from Kuroda Tetsuzan Sensei and referenced fluidity of movement especially. Renfield was dressed very fastidiously—a personalized black gi with black mon stitched on

Renfield Kuroda on left and Bill Henderson on right, demonstrating Mugai Kumitachi while students Johnathan Pierson, Igor A. Panteleyev, Jason Hatcher and Farid Sheikh watch

The Senpokan is a very nice little dojo nestled in an industrial sector in Costa Mesa, CA. The walls are lined with many bokuto and other wooden weaponry, There is a huge wooden oni mask hanging on the wall that Alverez Sensei brought back from Hataya

both front and back. His blade had a healthy amount of sori and he moved effortlessly as he instructed.

The next morning when we arrived at Senpokan, Niina Gosoke started the day with training of the mind.

Everyone gathered with their pens and paper for note-taking and the lessons that followed.

Soke began in explanation of blade positioning while cutting. He discussed tenouchi, addressing how in Mugai Ryu we grip Tsuka with the bone of the thumb more than the web between our thumb and forefinger. He further explained the web acts as a shock absorber. The other way of gripping, he stated, provides maximum power to be transferred into monouchi of the blade. He spoke about maintaining level height while moving, how not to telegraph a draw by pulling the saya back before drawing forward, and so much more.

It seemed as though our heads could swell no more from instruction—when we began to work on what evolved into a very long day of kata. With so many people in the Dojo, student groups had to cycle.

Performing all training elements—three to four times each—before stepping off the tatami mats to allow the next group of students their opportunities with their swords we realized the high-quality padded synthetic tatami flooring made performing kata from seiza comfortable enough to not have to wear knee pads. We all realized there was no need for them, which added tremendously to the experience.

The Senpokan Dojo, nestled in an industrial sector, was extremely nice. The walls were lined with bokuto and wooden weaponry. A huge wooden oni mask hung on the wall, which we learned Alverez Sensei brought from Hataya Sensei's shop in Japan. A stage was fashioned in front at shomen under the kamidana and featured a full set of yori along with a war fan, daisho, and incense burning station.

The Senpokan Dojo definitely had a traditional Samurai Dojo appeal. There were photos off to the side of Nakamura Taizaburo Sensei and it was comforting to see his picture displayed there.

As day continued into evening, we worked through the Mugai curriculum knowing testing would soon follow, but before testing occurred, Alverez Sensei required our assistance to roll mats for an upcoming Tai Kai the following weekend.

When I say he needed help, I mean he required an army of mat rolling experts. I do not recollect the

Renfield Kuroda performing Sha from Goyo no kata

Niina Gosoke of the Meishiha Mugai Ryu, preparing for the lessons of the mind, while Renfield Sensei and Bill Henderson watch on.

Soke began in explanation of blade positioning while cutting. He discussed tenouchi, addressing how in Mugai Ryu we grip Tsuka with the bone of the thumb more than the web between our thumb and forefinger. He further explained the web acts as a shock absorber. The other way of gripping, he stated, provides maximum power to be transferred into monouchi of the blade.

exact number of mats, but it nearly two hundred or so. All Swordsman and Swordswomen know, this builds character and everyone joined in to get it the job done so a lot of character was achieved.

When testing took place, Kurt and I from the Virginia Kenshinkai tested alongside 12 others who were working towards Shodan to Yondan in ranking. Testing was conducted in groups staged in front of Renfield Sensei and Alvarez Sensei.

Each of us performed 2 kneeling kata and 3 standing with reiho before and after for each test. Afterwards, a shizan (cutting of tatami) session was required.

Before we hit the floor for kata, we each recited "some *beautiful Japanese*" as Renfield Sensei put it. In Mugai Ryu there are poems we must memorize in order to advance.

It was a privileged for me to be able to test twice for both shodan and nidan and this gave me a lot of floor time and allowed me demonstrate what I had learned from Dave Drawdy Sensei over the years.

The shizan for the shodan test was sayu kesa on a single mat, for nidan it was kesa, gyaku one side and kesa gyaku on the other side, followed by a single kesa futomaki. As the tests progressed the futomaki got bigger and bigger—up to four mats. I hadn't seen a four mat roll until this time—it dwarfed the cutting stand.

Also, for higher-ranked tests there were draw kiriage cuts both from standing and seiza. It was tough on the practitioners testing for higher ranks, but all-in-all everyone did their best. Some students were told they should take the upcoming cutting seminars coming to make certain they would be on point with their level once promoted.

We are fortunate to have our seniors in the Sword Art community instructing us in order to achieve our goals and aspirations related to the Mugai Ryu.

To wrap up this glimpse into the Mugai Ryu Gasshuku 2010 now passed, I want to congratulate Kurt from our home dojo of the Virginia Kenshinkai, on achieving shodan

Renfield Kuroda Sensei on the left working on Kenjutsu techniques with Bill Henderson on the right.

Renfield Kuroda Sensei looks on while students attempt Sha from Goyo no kata

Renfield Sensei provided most of the teaching during our two days of training. Highly insightful, he regarded every movement with a watchful eye resulting in our working together in a classic manner. And, he demonstrated Kata, which he stated do not work the way they are configured. He delivered in-depth explanations regarding how slight adjustments accomplish Swordsmanship ideals.

The photography in this article was provided courtesy of Mr. Richmond

(currently a nidan).

For me . . . trips such as this can be fiscally challenging, especially with a young family and the responsibilities of parenthood, but getting to attend and learn from Sword Art leadership is a penny well spent. And, oftentimes the family can travel along to enjoy these events as well. I'm blessed to have a supportive wife and son already practicing some forms at the tender age of four.

Certainly, building a relationship within the Sword Arts Community is extremely important to each of us—it is good to experience the bigger picture in our arts—this knowledge reinforces our desire to continue and to learn.

Towards the end of Gasshuku, ceremonies took place. Okuden Certificates were dispersed. These were seemingly hidden keys to hidden doors of technique for the higher students such as Bill Henderson, Brian Pettit, Dave Neeley and Rob Rivers.

After these ceremonies, Big Tony Alvarez Sensei explained he has his own personal kamon, which was given to him by a former Sensei of another art form he practices.

His Senpokan Dojo is represented by this kamon as well. He explained students who reach certain levels in the Arts—by both rank and personal sacrifices made to keep the arts alive, receive special endowments with crested kimonos. These Seniors represent Alvarez Sensei in the states and in Japan as well.

He conferred one such kimono to a student by the name of David Creger—one of the Senpokan Dojo students.

While I do not know what hardships Creger may have experienced to compliment his training, I do understand hardship comes hand-in-hand with our training—whether Eastern or Western Sword Arts.

It can be tough to withstand training, which can be as demanding as our love for Sword Arts can be. This is especially true, when the demands of daily life are entangled as we sacrifice in training, practice and travel. Rewards for the Sword Arts Practitioner often come from within.

Creger must have experienced his own trial and tribulations in the demonstration of his dedication to Big Tony Alvarez Sensei. And, we are assured his was a much deserved reward from a forthright Sensei—everyone cheered Creger on as he was instructed to wear his kimono for Tai Kai, which followed.

The Mugai Gasshuku and our journey of discovery I believe was worthy to share with Katsujinken, a Sword Arts Journal readership because this provides an understanding and accounting for our experiences at remote training and testing sites.

And, at Katsujinken, we invite "our readers" to share your memories and experiences as well.

Jason Lee A. Hatcher
Editor-in-Chief Katsujinken

2011 Swordfest, A Pictorial Essay

Swordfest is an annual event held by the Virginia Kenshinkai at Capital Area Budokai in Alexandria Virginia (Metropolitan Washington Region of the United States of America).

Attendees are welcomed each year to wear traditional dress at this "indoor" (rain or shine) event. The Virginia Kenshinkai welcomes photographers, sponsors and vendors each year as well—to capture National and International Sword Artisans doing what they love most.

These "STARS" of both Eastern and Western Mixed Edged Weaponry are anxious to demo for guests who are comprised of Sword Art and Non-Sword Art Community members—making this an opportunity to educate and inform individuals not necessarily accustomed to authentic Sword Arts of "any" tradition.

Swordfest features crafts, trades people and merchants as well as good food and all around good people. As a member of the Budokai—now a 501 (c)(3)—the Virginia Kenshinkai promotes appreciation of Sword Art culture and history.

Fran Vall, Sensei—Swordfest 2011

Sword Artisans, Instructors and Sensei Exhibitors—2011 Swordfest	
Hobyokan Dojo	David Hall Sensei
Large Two-handed Sword	Stephen Hicks Instructor
Nakamura Ryu Happo	Dave Drawdy Sensei
New York Historical Fencing Assoc.	Michael Edelson Principal Instructor
NoVA Kyudo Kai	Bill Reid Sensei
NoVA Tendokan, Aratashii Naginata	Fran Vall Sensei; Raymond Sosnowski Sempai
NoVA Assalto, Spadone	Steven and Kara Reich Instructors
NoVA Budokai Kendo	Brian Sherry Sensei
Sekiguchi Ryu Kenjutsu	Dan Keupp Sensei
Shakuhachi Flute	Raymond Sosnowski Sensei
Shin Dojo, Mugai Ryū Meishiha	Rob Rivers Sensei
Shin Shin Sekiguchi Ryū	Dan Keupp Sensei
Shutokukan Dojo	Meik Skoss Sensei
Toyama Ryu US Battodo Federation	Sang Kim Sensei
Yagyu Shinkage Ryū Hyoho	Paul Manogue Sensei

Per Boylan of Mugendo Budogu vendors at Swordfest 2011

Katsujinken, a Sword Arts Journal is happy to provide this pictorial essay of the 2011 Swordfest, Instructors, Sensei and Sempai along with their student performance artisans and the guest adventurers willing to engage in elementary training sessions.

With a suite of volunteer photographers joining the 2011 Swordfest—images exceed 3500 photographs and we cannot include them all. So, the Katsujinken Team selected a series of images to include for your enjoyment . . .

We have decided to present our compilation of imagery in a collage; understandably, some photos will be more difficult to view than others—but we wanted to include as many Swordfest memories as possible for reader enjoyment.

Katsujinken Team

UBBC
U.S.A
フラン
FRAN

CAPITAL AREA BUDOKAI

NYHFA

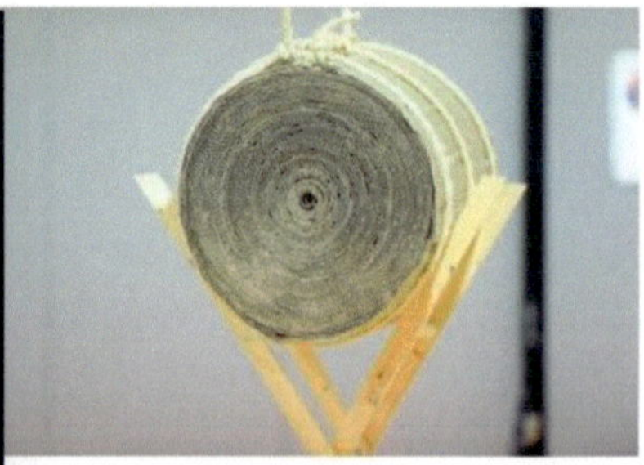

All images contained within the 2011 Swordfest Pictorial Essay are compliments of the *Katsujinken, a Sword Arts Journal's* photography friends:

DEOPAGES.COM

David Zlotin—DEO Studio and provided Katsujinken, a Sword Arts Journal with unique angulated images and artistic renderings.

Jim Blocker—Middle Blocker Photography who provided excellent high-density screen-sized photographs.

Tom and Brenda Turk—Piratical Photography these folks totally rocked Swordfest with a professional still photo set as well as their sports arena staged imagery.

Honorable mentions are extended to:

⇒ Bangaley Cisse—Bangaley Photos
⇒ Seon Corbin—Corbin Photography
⇒ D. Owens—CSM Imaging, LLC
⇒ Jermaine Dent, Day 1 Productions
⇒ Ron Johnson, Sensei—Johnson Digital Photography
⇒ Ken Howard—Ken Howard Images
⇒ Henry Maynard—Maynard Photography
⇒ Dan "Nubs" Sugre—Nubs Photography
⇒ Charles House—Pulp Added Words
⇒ Bobak Shafiel—Shafiel Photography
⇒ Sarah Sertic—Tribal Spider Arts
⇒ Zagros Photos

Images from our suite of mentions we were unable to include in the 2011 Swordfest, a Pictorial Essay—but Katsujinken extends a hearty thanks to "all" photographers who took time to join in the festivities and capture this event.

Our readers can reach out to DEO Studio, Middle Blocker Photography and Piratical Photography to request varied artistic renderings for gift-giving and | or private collections.

At Katsujinken Magazine, in our quarterly releases, we feature a Dojo or Training Hall SPOTLIGHT within each issue. Oftentimes, places where training occurs in the Martial Arts Community, tend to be a second home for the many practitioners who train there.

These Training Halls often are quite unique in their own right. While some Dojos seem basic—housing only space to train in—many are not, regardless each of them hold a position of merit for their family of practitioners.

When Training Halls push the envelope of accommodation—originality results. These Dojos often incorporate living quarters (with showers and kitchenettes), art galleries of sorts, weaponry arsenals (some could supply small armies), and retail and vintage shops; in addition to huge matted areas where Shrines are poised for reflection and dedicated to the Gods or Goddesses of the Sword Art Forms. The simplicity of each Dojo—when coupled with their integral use of every square inch of available space—results in truly exceptional museumesque environs where culture and training meld together creating the synthesis, which becomes greater than the measure of floor space rendered.

For our first issue of *Katsujinken, a Sword Arts Journal,* we feature Robert Zingg Sensei's Karate Center Dojo located at 111 West Second Avenue, Suite B, Ranson, West Virginia 25438. We regard his venue as the . . . *Welcoming Cave of Weapons*. . .

Zingg Sensei's Karate Center Dojo is nestled amidst the rustic town of Ranson, West Virginia—just outside of historic Harpers Ferry—an infamous Civil War Era historical site. When Katsujinken visited Zingg's Dojo we found the a most wondrous cave—but this is definitely not an "ordinary cave."

You don't find foliage growing in abundance here, nor wild animals roaming free; however, you are greeted by a kick-butt yellow belted bear holding a "Welcome" sign when entering this Sword Arts dwelling. Once you traverse the front door of Zingg's Dojo you pass through a small supply shop and then step onto bright red carpets, which lead to the inner sanctums of his Cave of Weapons.

Once inside this cave, your breath is virtually taken away in review of the abundance of weaponry of all sorts, sizes, shapes and eras of Sword and Martial Art history. There's so much going on—it is difficult to capture it all in your memory banks upon just one visit.

So, Robert Zingg Sensei and the Zingg Dojo—this month's DOJO SPOTLIGHT is focused on you! Awesome job.

Katsujinken Magazine.

UFC
CHAMPIONNAT DU MONDE DE KENPO
A GENEVE
Bushikai

orlando tai-kai 2011
EVENT RESULTS & REVIEW

Hataya Sensei presented remarks concerning this season's Tai Kai and said . . . *"Shodans and Nidans testing this year, were among the best he has seen."* He stated, *"Sandans and above require some improvements,"* and shared, *"often teachers take the time to train students while neglecting their own personal training—continuously at the business of watching their students; however, teachers like their students must find time to practice too."*

Hataya Sensei followed, *"This is one of the best Tai Kai events he has attended in the United States of America."*

The Orlando Tai Kai presented many prizes—some of which were donated by MAS (Martial Art Swords) and Nihonzashi.

MAS generously donated an L-6 Katana for Overall First Place—Yusho winner.

Nihonzashi generously donated a Tetsuba War Club for First Place Dodan Cutting, two Gekken sparring swords to First Place Kumitachi Division and also three boxes of Tatami Omote for each division with First Place Tameshigiri.

For team cutting events, Wally Hostetter of Shining Moon, donated Certificates for Free Sword Sharpening.

Medals featured here next to the *2011 Orlando Tai Kai Winner's List* belong to Carl Jenkins Sensei (Gold—Overall Champion—Yusho) and Jason Hatcher (Bronze for 3rd place Nidan kata and Silver for 2nd place kumitachi Winner).

The Katsujinken Team

Orlando Tai Kai Hosting Sensei

2011 Orlando Tai Kai Winner's List

Event		First Place	Second Place	Third Place	Forth Place	Fifth Place
Batto Kata	Shodan and Under	Sergio Beltran	Kyle Kim	Nick Paperno		
	Nidan | Sandan	Henry Kim	Tracy Snell	Jason Hatcher		
	Yondan and Above	Carl Jenkins	Mike Soriero	Andy Seminick		
Tameshigiri	Shodan and Under	Mark Kurihara	Clay Wilson	Jeff Jones		
	Nidan | Sandan	Paul Koseenanonth	Henry Kim	Ron Duponte		
	Yondan and Above	Carl Jenkins	Hal Smith	John Boyer		
Wakizashi	All Levels	Carl Jenkins	Hal Smith	Adrian Demoret		
Dodan	All Levels	Hal Smith	Carl Jenkins	Jeff Jones	Andy Seminick	Mike Soriero
Kumitachi	All Levels	Kyle Kim	Carl Jenkins	Rob Crook		
		Henry Kim	Jason Hatcher	Russell Gordon		
Team Cutting	All Levels	Bob Elder	Jason Hatcher	Neal Earhart		
		Tom Smyth	Carl Jenkjins	Alex Odell		
		Sakaida Sensei	Dave Drawdy	Matt Odell		
Grand Champion	Carl Jenkins					

Katsujinken

A Sword Arts Journal

SNEAK PEEK

- ⇒ *Dave Drawdy Sensei, Nakamura Ryu Batto Do*—SENSEI SPOTLIGHT INTERVIEW
- ⇒ Dr. Manouchehr Moshtagh Khorasani with fellow colleagues—*Battle Formation, Tactics and Weaponry Choices described in The Book of Kings*, by Ferdowsi
- ⇒ Kagenori Ueno Soke Interview—*Rise of Tenshinsho Jigen Ryu in the United States of America*
- ⇒ *Samurai Speak*, by Kent Jensen
- ⇒ *International Batto-do Federation in the United States of America*—Words from Seniors
- ⇒ *A Brief Introduction into 16th Century Bolognese Swordsmanship*, by Steven Reich of NoVA Assalto
- ⇒ Group Photo Contest—Prize Features—with submission request for images, photographs and pictorial essays.
- ⇒ *The Medieval Art of Combat with Sword and Buckler Interpreting Europe's Oldest Fencing Treatise*, by Rowland Warchezha

Submissions Call: at *Katsujinken, a Sword Arts Journal* we publish academic works, articles, comics, dojo features, essays, events (calendar of competitions and their results), fiction, interviews, journalistic features, mythos behind the arts, photography, poetry, sensei profiles, and stories.

Surprise us; we often don't know what we'll like until we read it - we remain open to new ideas and talents.

For our March Issue we are featuring a ***"Group Photo Contest"*** with prizes! Be sure to remit your images to submit@katsujinkenmagazine.com—please include relative information regarding participants, event place and time, and ensure your group approves of their inclusion in your submissions.

For upcoming issues, we are always open to coming events as well as post-event results. We intend to expand the ***"Events Section of Katsujinken"*** in order to provide readers with information to aid in further development of their Sword Arts.

Both Eastern and Western Sword Art enthusiasts, practitioners and trainers as well as professors and authors are welcomed to submit works to share with the Sword Arts Community through Katsujinken—for both national and international readership.

Advertisers are welcome, please contact Jason Lee A. Hatcher at Jason@katsujinkenmagazine.com to discuss requirements.

If you are interested in a **Subscription** to *Katsujinken, a Sword Arts Journal* please contact michelle@katsujinkenmagazine.com.

The Katsujinken Family thanks you our readers as well as our phenomenal

A special note of thanks to Katsujinken's Publisher—DonnaInk Publications , L.L.C. for building the publishing platform.

DonnaInk Publications, L.L.C.
www.donnaink.org

Katsujinken

活人剣

www.katsujinkenmagazine.com

Made in the USA
Las Vegas, NV
03 November 2026

57422974R00038